The Divine Number 12

The Divine Number 12

CJ Dodaro

This is a self-published book.

First Edition

Published: January 2021

ISBN: 9798701272857

Other titles by CJ Dodaro:

Being Successful in Your Own Business - A Step-by-Step Guide to Success

(This is a (3) Book Series)

Book 1 of 3 in the Series: **Your Business Setup**

Book 2 of 3 in the Series: **Work on Your Website**

Book 3 of 3 in the Series: **Get Social, Videos, and Money Management**

Book 4: **Write it - Publish it - FREE!**

Book 5: **Make Money - Work at Home with an Unclaimed Money Recovery Business**

Book 6: **Make Money - Work at Home with a Tax Sale Overages Business**

Book 7: **Make Money - Work at Home with a Tax Lien Certificates & Tax Deeds Business**

Book 8: **How to Build a Deck - Step-by-Step Instructions**

Copyright

The Divine Number 12

CJ Dodaro

First Edition

ISBN: 9798701272857

© 2021 by CJ Dodaro

License Notes

<u>Dedication</u>

This book is dedicated to my wife, **Karen**,

who has always supported me in everything that I have tried to accomplish.

Table of Contents

The Divine Number 12

Introduction

Some of you will simply find the book entertaining and a fun read. Others will take it to heart, and attempt to incorporate the **number 12** in their lives whenever they can, in order to benefit from it. I hope you are of the second batch of individuals, for I only wish you good things in this life.

Numbers are everywhere throughout history and in every aspect of our present society. There are lucky numbers, unlucky numbers, positive numbers, negative numbers, etc.

This book is all about the **Divine Number 12**. As you will learn here in, this number is present everywhere throughout the human race, both in our history, as well as, in our present society. It is in all of our important religions, sciences, arts, and even in other abstract areas of our lives.

It is a very important number, signifying: harmony, peace, success, and self-improvement. Let's take a look at where this number is found, and how it ultimately can affect our lives.

Chapter 1: <u>The Invention of Numbers</u>

Who actually invented numbers is a much heated debate. Not surprising, as many people would like their own culture to be responsible for such an important development in human history.

It should be noted here, that there is a difference between numbers and a numbering system. For example, the **Arabic** numeral system we are all familiar with today is usually credited to two mathematicians from ancient **India**: **Brahmagupta** from the **6ᵗʰ century B.C.** and **Aryabhat** from the **5ᵗʰ century B.C.**

Historians believe numbers and counting expanded beyond one around **4,000 B.C.** in **Sumeria**, which was located in southern **Mesopotamia**, in what is now southern **Iraq**. One of the first civilizations to feature cities that were centers of trade, the people of **Sumeria** needed new methods of counting and record-keeping.

Perhaps the oldest mathematical artifact in existence, the **Ishango Bone** was unearthed in **1950** in the then **Belgian** colony of the **Congo** (now the **Democratic Republic of Congo**). The bone, probably a fibula of a baboon, large cat, or other large mammal, has been dated to the **Upper Paleolithic Period** of human history, approximately **20,000-25,000** years ago. It is **10 cm** long and bears an articulated, organized series of notches readily identifying it, to many observers, as a tally stick. However, its original purpose remains the subject of debate.

However, all that being said, the first human may have simply realized that he had **5 fingers** on each hand. This may have led to him being able to count, but obviously, no one knows for sure.

<u>Oldest Known Civilizations</u>

History is constantly changing as new discoveries are being made. The things that we believed yesterday may not be so true today. Here is a little history on civilizations:

The **Ancient Egyptian Civilization** was established when **King Menes**, the first pharaoh, unified **Upper and Lower Egypt in 3150 BC.**

Extending from modern day **Afghanistan and Pakistan** to northwest **India**, the **Indus Valley Civilization** covered **1.25 million kilometers**, making it the most widespread civilization of the ancient world. The earliest people gathered around the basin of the **Indus River**, establishing farming settlements. **3300 BC** is when historians generally clock the first signs of urbanization.

The **Maya Civilization** was largely made up of the indigenous people of **Central America and Mexico**. Their hunter-gatherer lifestyle can be traced back to **7000 BC**, but the first permanent villages were built around **2600 BC.**

The **Jiahu settlement**, which is located in the central plain of ancient **China**, is an area known today as the **Henan Province**. The people of this settlement belonged to the country's oldest recognized civilization, around **7000 BC.**

For a long time, scholars believed that **Mesopotamia** was the first civilization. Located between the **Tigris and Euphrates** rivers, its name means "between (*meso*) rivers (*potamos*)." Today, the region encompasses **Iraq, Kuwait, Turkey, and Syria**. For thousands of years, the early people lived in small settlements which eventually transformed into a scattering of farming communities around **8000 BCE.**

Although it is still unclear when the first stone-based structures were erected, recent radiocarbon dating studies conducted in **Göbekli Tepe**, **Turkey** indicate that the construction of the iconic T-shaped limestone pillars stretches back more than **12,000** years.

Although the people who settled in **Mesopotamia** are often credited as the first civilization, new research shows that **Aboriginal Australians** are the oldest known civilization on **Earth**. The **Aborigines** can trace their ancestries back to about **75,000 years ago**, but became a distinct genetic group around **50,000 years ago**.

In early civilization, **twelve** became a very important number to mathematicians and astronomers. **Twelve** is a number that is divisible by 2, 3, 4 and 6 as well as by itself and 1. That's two more numbers than can be divided equally into 10. Astronomers divided a year, or the average time it takes the **Earth** to make a complete revolution around the **Sun**, into **12** months. **Days** (the time it takes the **Earth** to make one revolution) were divided into 24 hours each with two definable parts (day and night) lasting roughly **12** hours each during spring and fall equinoxes.

So, there you have it; **History** as we know it today. Who knows what tomorrow will bring.

Chapter 2: <u>Numbers are Everywhere</u>

I can't even imagine a life without numbers. We use numbers in almost all phases of life. We count our life in years. We count each day in minutes and hours. We count the months by days. We count the years by months and the centuries by years.

Numbers are used in our height and weight measurement. We even count calories by numbers.

We use numbers in all types of construction measurements. There are numbers in clinical measurements.

Numbers are found in **History** to show dates of special events. In **Science**, numbers are used in many different equations to calculate distance, size, etc. Numbers are used in **Shop** class, whether in woodshop to measure boards or metal shop to also make measurements. In **Home Economics**, numbers are used in ingredient measurements. Even in **Gym Class** numbers are used to measure distance and the number of repetitions of a particular exercise and obviously in **Mathematics**, where numbers are used in all types of calculations.

Numbers are truly everywhere. But what makes a particular number special? It is how often that number appears in life itself and what that number represents. Before we explore the **number 12**, let's take a look at some other special numbers.

<u>The Number 3</u>

Three is the smallest **number** we need to create a pattern, the perfect combination of brevity and rhythm. It's a principle captured neatly in the Latin phrase omne trium perfectum: everything that comes in threes is perfect, or, every set of **three** is complete.

Total completeness, inner peace and sanctity are denoted by the number **three**. There may be **12 Apostles of Jesus**, but of them, **three remain his favorites**. Besides, **Satan** tempts **Jesus thrice** before giving up, while it was on the **third day of Creation that Earth was created**. Other than these, there remain several other uses of the number throughout the **Holy Bible**.

The **number 3** is one of the most **important numbers** through **Bible**, it is a **number** of harmony, of **God's** presence and of completeness. There are **3 to the Holy Trinity**: **God the Father, God the Son, and the Holy Spirit**. **Number 3** is the **number** of eternal life, as **Jesus** was resurrected after **three** days of being dead. **Three** people saw the **Holy Transfiguration of Jesus**; they were **Peter, John and James**.

There were **three** patriarchs before the flood: **Abel, Enoch and Noah**, and three patriarchs after the flood: **Abraham, Isaac, and Jacob**. The **New Testament** has **27 books**, which is 3^3 **(3*3*3)**, and it is largely connected with **Jesus**.

Jesus prayed **three** times before being arrested, he was put at the cross at the **third** hour of the day, there were **3** hours of darkness that covered the land while **Jesus** was suffering on the cross and died at the ninth hour which is **3pm**. There were **three** people crucified that day; **Jesus** and two others.

The **number three** can be translated as the symbol of **God's** power. For example, only **three** people had the permission to ask **God** any questions: **Solomon, Ahaz and Jesus Christ**. **God** gifted **three** things to **Israel**: the law of **God**, the land of **Israel** and the world to come. **God** himself is described in a **triple** phrase construction, which is very common in the **Bible**: "which is, and which was, and which is to come" (**Revelation 1:4**). **God's** throne is located at the "**third heaven**".

Three is the first number to which the meaning 'all' was given. It is the **Triad**, being the number of the whole as it contains the beginning, a middle and an end. The power of **three** is universal and is found in the nature of the world as heaven, earth, and waters. It is found in us humans as body, soul and spirit.

The **number three** can be interpreted in many different senses: spirit, mind, body; in a circle of synthesis: past, present, future; enclosed in the ring of eternity: art, science, and religion bound in a circle of culture.

Three is the number of sides found in a triangle, which is the shape of the pyramids.

Three represents the triad of family: male, female, and child. It is also used to describe our life cycle: birth, life, and death.

Only **three** primary colors are needed to mix most other colors; red, yellow and blue.

While there are many, many, more examples of the **number 3** in our world, let's explore another great number; the **number 7**.

The Number 7

The **number 7** is another important number. **God** created the world in **6 days**, and on the **seventh** day, he rested.

There are **7** days in a week (Don't ask me why our calendars start on the day of rest; **Sunday**.).

There are **7** oceans, **7** continents, **7** vertebrae in the neck, **7** layers of skin (2 outer and 5 inner), ocean waves roll in **sevens**, the rainbow has **7** colors, sound has **7** notes, and there are **7** directions. The **Jewish Menorah** has **7** candles, there are **7** holes in your head (go ahead and count them), a cube has **7** dimensions (including the inside), the male body has **7** parts, and the **number 7** is used **735** times in the **Bible**.

There are **7** deadly sins, **7** virtues, **7** gifts of the **Holy Spirit**, **7** classical planets, **7** numbers in a phone number (after the area code), **7** hills in **Istanbul, Rome and Jerusalem**, **7** liberal arts, **7** wonders of the ancient world, **7** is the number of games in the playoffs for **NHL, MLB and NBA**. The number **7** is also important in **Hinduism, Islam and Judaism**.

While there are other important numbers in the world, I do not have the time to go over all of them in this book, which is about **The Divine Number Twelve**. So, let's explore the **number 12**.

Chapter 3: <u>The Number 12</u>

The first thing you should do if you want to discover the meaning of **number 12** is to discover the meaning of its components. It means that you should try to discover what numbers 1 and 2 mean.

Both of these numbers are angel numbers and their symbolism is very powerful. Number 1 is known as a symbol of self-leadership and authority.

This number also indicates new beginnings and positive changes. If number 1 has been sent to you, it probably means that something new will take place in your life and you should be ready for it. You shouldn't have any fears, because you have the celestial and divine protection. We can also say that number 1 is considered to be a symbol of purity and positive thoughts.

When it comes to number 2, we have to say that this number is a symbol of peace and balance, as well as a symbol of your soul destiny.

Number 2 will make your faith stronger and it will help you accomplish your soul mission. Having this number by your side means that you will have the complete trust in your guardian angels.

The **number 12** is made up of two numbers, 1 and 2. 1 is a prime number and signifies the beginning and the singular nature of the universe. It is also related to completion, perfection, harmony, motivation, achievement and independence. 2, on the other hand, is all about seeing two sides of any situation, diplomacy, partnership and the mutable nature of life.

Number 12 is made up of the vibrations of numbers 1 and 2. It is clear that **number 12** will motivate you to take action and to take an important step in your life. Also, they will bring order into your life and you will be ready to experience the positive changes that will happen.

Very often **number 12** is considered to be a symbol of harmony and peace. If you are seeing this number very often, it means that your life will be peaceful in the future. Another meaning related to **number 12** is self-improvement. It indicates that this number will help you be successful and make improvement in all areas of your life.

Twelve has become an important number in most cultures, and therefore it occupies an important place in the brains of most humans.

<u>In Astronomy</u>

This **Divine Number** is approximately the number of full lunations of the moon in a year. It is also the number of years for a full cycle of **Jupiter**, which was historically considered the brightest 'wandering star'.

<u>In Measurements</u>

We inherited the **English** measurement system here in **America**, which was based on using the parts of a person's body. The average length of the distance from the top of the index finger to the first joint was equal to an inch. The rough size of a man's foot was equal to **12** inches. Although some people look at the **English** measurement system as backward and archaic, they were actually quite creative, and in many ways, make a lot of sense.

In Time Keeping

It is central to many timekeeping systems, such as the **Western Calendar** and the units of time of day. There are **12** months in a year. Also, there are **12** hours of day and **12** hours of night. Many of us eat lunch at **12**-noon or stay awake until **12**-midnight. **Twelve** has become a very important number in most cultures, and therefore it occupies an important place in the brains of most humans.

Mathematical Properties

In **Mathematics**, **Twelve** is a composite number. It is also the smallest number with exactly six divisors, with its divisors being: 1, 2, 3, 4, 6, and **12**. **Twelve** is a highly composite number, with the next number being twenty-four.

Twelve is also the smallest abundant number, being the smallest integer for which the sum of all of its proper divisors is greater than itself $(1 + 2 + 3 + 4 + 6 = 16)$. **Twelve** has a perfect number of divisors, causing it to be a sublime number. The sum of its divisors is also a perfect number. Excluding the number 4, and adding the remaining subset of **12**'s proper divisors that add up to **12**, causes 12 to also be considered a semi-perfect number.

A dodecagon is a **twelve**-sided polygon. A dodecahedron is a **twelve**-faced polyhedron. Octahedrons and cubes both have **12** edges. Regular icosahedrons have **12** vertices. **Twelve** is also a pentagonal number.

The **Kepler Conjecture**, an accepted theorem, proves that the densest 3-dimensional lattice sphere packing has each sphere touching **12** others. This is almost certainly true for any sphere arrangement. In 3-dimensions, **Twelve** is also the kissing number.

The smallest weight for which a cusp form exists is **Twelve**. Given by the **Ramanujan t-function** and which is (up to a constant multiplier) the 24th power of the **Dedekind eta function**, this cusp form is the discriminant $\Delta(q)$, whose **Fourier coefficients** are given by the **Ramanujan t-function**.

This fact is related to a constellation of interesting appearances of the number **twelve** in mathematics ranging from the fact that the abelianization of **SL (2,Z)** has **twelve** elements, the value of the **Riemann zeta function** at -1 i.e. $\zeta(-1) = -1/\mathbf{12}$, and even the properties of lattice polygons.

There are **12**, 3 x 3 **Latin squares**.

There are **twelve** cubic distance-transitive graphs, as well as, **twelve Jacobian elliptic functions.**

Probably originating in **Mesopotamia**, is the duodecimal system ($\mathbf{12}_{10}$ [**twelve**] = $10_\mathbf{12}$), which is the use of **12** as a division factor for many ancient and medieval weights and measures, including hours.

Twelve is represented as **C** in base thirteen and higher bases, such as hexadecimal. In **base 10**, the **number 12** is a **Harshad number**.

Twelve is a **Pell number**, as well as, a **pronic number**.

Chapter 4: <u>The Number 12 in the Bible</u>

Most numbers have special meanings in the **Bible**. The **number 12** is considered to be one of the most important numbers that are mentioned in the **Bible**. Found **187 times** in the **Bible**, for **Christians**, the **number 12** typically means perfection or authority and is often used in a context of government. For instance, because **Jacob** has **12** sons, they form the **12** tribes of **Israel**. There are also **12** minor prophets that dot the text of the **Old Testament**.

Actually, each of the tribes of **Israel** had **12,000** of **God's** servants and we have already said that there were **12** tribes of **Israel**.

We have also to say that the **Bible** contains **12** books of history.

There were **12** administrators in the kingdom of **Solomon**.

In the **Old Testament Tabernacle**, a mobile temple of sorts, the priests were to place **12** unleavened cakes (**Leviticus 24:5**). The unleavened nature of these cakes may have symbolized the same bread used in **Passover**, when the **Israelites** left **Egypt**. Some other elements of **12** in the **Tabernacle** included **12** silver plates, bowls, bulls, rams, and male lambs (**Numbers 7**).

Jacob isn't the only descendant of **Abraham** to have **12** princes (tribes) come from him. **Ishmael, Abraham's** son he bore through his servant **Hagar**, also had **12** princes come from his line (**Genesis 17:20**). These **12** tribes of Ishmael later played a role in the history of **Israel**.

When the **Israelites** wandered in the desert after they fled **Egypt**, they had **12** spies go and scout the **Promised Land** (**Numbers 13**).

Another biblical fact is that **Elijah** built the altar that was made up of **12** stones.

We see **Jesus** chose **12** disciples, and the disciples later replaced **Judas Iscariot** with **Matthias** to keep the number at **12** (**Acts 1**), before the **Holy Ghost** was delivered to the apostles, showing the perfection and authority of those who followed **Jesus**.

Jesus first spoke in the temple with religious leaders at the age of **12** when he and his family had traveled to **Jerusalem** for **Passover** (**Luke 2:41-52**). At this age **Jesus** was separated from his parents. The religious leaders marveled at his wisdom.

Another fact mentioned in the **Bible** is that the Virgin Mary stayed **12** years in the temple. It is also written in the **Bible** that **God** had **144,000** servants, which actually represents the multiplication of numbers **12** and **12,000** (**12 x 12 000** gives us 144,000).

When **Nebuchadnezzar** has a dream that is interpreted that he will act like a wild beast for a while, the dream ends up fulfilled **12** months later (**Daniel 4**).

When **Jesus** performs the miracle of the feeding of the **5,000**, the disciples gather **12** baskets full of leftovers (**Matthew 15**).

Solomon appointed **12** district governors over **Israel**, showing how **12** has symbolic importance in government as well (**1 Kings 4:26**).

Elisha, a prophet, while plowing **12** oxen is called into his prophetic ministry by **Elijah** (**1 Kings 19:19**).

When **Israel** returns after their captivity, **Ezra** sets apart **12** priests (**Ezra 8**).

A tree of life will also be in the new world **God** brings at the end of time (**Revelation 22**) with **12** fruits, one for each month of the year.

In the **Book of Revelation** it is said that there will be **12** gates in **God's** kingdom and there will also be **12** angels who guard the gates. It should also be noted that the number **12** appears **12** times in the **Book of Revelations**.

It is interesting that each gate got the name of a certain tribe of **Israel**. It is also written in the **Bible** that there were **12** beautiful stones that would be used as the **New Jerusalem's** foundation.

These are only some of the facts related to number **12** that are appearing in the **Bible**, but there are also many others. We can say that the number **12** is considered to be a symbol of perfection and also a symbol of government.

It could be also perceived as a symbol of completeness and authority. We all know that number 666 is used as a symbol of devil, so we can say that the number **12** is the opposite number to number 666, while **12** is a symbol of **God**. Another fact is that 666 is the number of hell, while number **12** always represents heaven.

In most cases, the number **12** is actually a representation of authority and perfection.

This number can also represent the church and faith in general. It can be also used as a symbol of divine rule, actually the symbol of the perfect government of **God**. Many people who are following the **Bible** prophecy think that number **12** could symbolize the return of **Jesus Christ** on the **Earth**.

It is clear now that this number is very important in the **Bible** and prophetically. It seems that this number was very important to **God**.

Chapter 5: <u>The Number 12 in Numerology</u>

Astrologers identified **12** major star constellations in the sky and created the **twelve** houses of the **Zodiac**. The number **12** is an important one when it comes to numerology since it signifies completion. It also belongs to the star sign **Pisces** who is known to be a spiritual sign that is in constant touch with the energies of the universe. If you have a birthday that falls on the **12th** or were born in **December**, you already have angel guides on your side, offering you much-needed support and advice should you need it.

12 is at the very end of the numerology spectrum, and it offers those who see it in their daily life the opportunity to turn over a new leaf by giving them a chance to wrap up a certain life stage and situation before moving forward to bigger and better things. This number is like a curtain call that allows you to get your affairs together so you can benefit from the windfall that the universe is about to bestow on you.

If you are a number **12** person, people know you as a dynamic and energetic individual who isn't afraid of pushing the boundaries. In addition, you are often at the center of action, preferring to be around people as well as assuming a leadership position. People come to you for advice and value you for your complete honesty. That being said, you are all about partnership and prefer to have someone by your side not only to give you much needed moral support, but also to provide you with an anchor in reality so you don't go off on an illogical tangent from time to time.

If you have an abundance of **12**'s in your life, you are thought of as a consumer as well as someone who places great importance on materialism. You like lavish surroundings, and you deserve the best. You also understand the dual nature of the world, and believe that happiness and suffering serve a certain purpose in everyone's life. Due to this kind of thinking, you are more likely to have the stamina to go through life in an easy and effortless manner, understanding and accepting of the various challenges you may encounter.

Number 12 people are also very spiritual, and are more likely to be co-creators of their karma because of the fact that they feel a very strong connection to energies around them. This may make them more likely to be selfless helpers, spiritual teachers or people to look to for moral and spiritual advice. Some may even sacrifice their beliefs and creature comforts just to make sure that they reach a certain level of spiritual enlightenment. When you have **12** on your side, you can be assured of a rich and fulfilling life through the ages.

Chapter 6: <u>The Number 12 in Government, In Nature, & Timekeeping</u>

<u>The Number 12 in Government</u>

The number of **Circuit Courts of Appeal** under the **U.S. Supreme Court**.

The number of jurors serving in a jury in the **U.S. justice system** (usually with 4 backup jurors).

The number of districts controlled by regional **U.S. Federal Reserve Banks**. The combined number and letter (e.g. **E5**) are printed in black below the serial number on any **U.S.** money bill to show where the bill was issued.

A1: Boston (Maine, Massachusetts, New Hampshire, Rhode Island, Vermont and most of Connecticut)

B2: New York (New York, northern New Jersey, Fairfield County, Connecticut)

C3: Philadelphia (Delaware, southern New Jersey, eastern Pennsylvania)

D4: Cleveland (Ohio, eastern Kentucky, western Pennsylvania, northern West Virginia)

E5: Richmond (Maryland, Virginia, North Carolina, South Carolina, southern West Virginia, Washington D. C.)

F6: Atlanta (Alabama, Florida, Georgia, eastern Tennessee, southern Louisiana, southern Mississippi)

G7: Chicago (Iowa, northern Indiana, northern Illinois, southern Michigan, southern Wisconsin)

H8: St. Louis (Arkansas, southern Illinois, southern Indiana, western Kentucky, northern Mississippi, eastern Missouri, western Tennessee)

I9: Minneapolis (Minnesota, Montana, North Dakota, South Dakota, upper Michigan, northern Wisconsin)

J10: Kansas City (Colorado, Kansas, Nebraska, Oklahoma, Wyoming, western Missouri, northern New Mexico)

K11: Dallas (Texas, northern Louisiana, southern New Mexico) and

L12: San Francisco (Alaska, Arizona, California, Hawaii, Idaho, Nevada, Oregon, Utah, Washington).

North Carolina was the 12th State to ratify the U.S. Constitution to join the Union, on 21 November 1789 (21 being the reverse of 12).

The 12th U.S. President: Zachary Taylor (Whig, Virginia, 1849-1850). His Vice President, Millard Fillmore (New York), was the 12th U.S. Vice President.

The number of years the U.S. President Franklin Delano Roosevelt served (1933 to 12 April 1945).

The speed limit (miles/hour) set by the first automotive law in the state of Connecticut, in 1901 (WOW!).

In Nature

Twelve is the number of years for a full cycle of **Jupiter**, which is the brightest of the ancient 'wandering stars'. It is also the number of full lunations in a solar year, as well as, the number of months in a solar calendar. It is the number of signs in the **Chinese and Western Zodiac**.

Timekeeping

12 lunar months equals a lunar year and adding 11 or **12** days completes the solar year.

Solar and/or lunar calendar systems have **twelve** months in a year.

Earthly Branches is a **12**-year cycle for time-reckoning used by the **Chinese**.

Evenly divisible by **twelve** into smaller units, are the basic units of time: 60 seconds, 60 minutes, and 24 hours.

There are **twelve hours** in a half day (a.m. and p.m.), with **12:00 p.m.** being noon or midday, and **12:00 a.m.** being midnight.

Chapter 7: The Number 12 in Science, Sports, & Technology

In Science

In the periodic table, the atomic number of magnesium is **twelve**.

The **Standard Model of particle physics** identifies **twelve** types of elementary fermions.

There are **twelve** cranial nerves in the human body.

The first part of the small intestine is known as the duodenum, and is about **twelve** inches long.

In **German** and in **Dutch**, the name for the duodenum means: 'twelve-finger bowel', and this section of the intestine was measured in fingerwidths, not in inches.

On the **Beaufort wind force scale, Force 12** corresponds to the maximum wind speed of a hurricane.

In Sports

In both **American Football and Soccer**, the number **12** can be a symbolic reference to the **Fans**, due to their support for the **11** players on the field. At **Texas A& M University**, they reserve the **number 12** for a walk-on player who represents the original **12th Man**, who is a fan who was asked to play when the team's reserves were low in a college **American Football** game in **1922**. Reserved for their supporters, the following do not allow field players to wear the number **12** on their jersey: **Hammarby, Bayern Munich, Atlético Mineiro, Feyenoord, Flamengo, Portsmouth, Seattle Seahawks, and Cork City**.

12 is the maximum number of players that can be on the field of play for each team at any time in **Canadian Football**.

One of the starting second-row forwards wears the number **12** jersey in most competitions in a **Rugby League**, with the exception of the **Super League**, which uses static squad numbering.

One of the starting centers, most often but not always the inside center, wears the **12** shirt in **Rugby Union**.

Each team has **12** players on the field at any given time in **Women's Lacrosse**, except in penalty situations.

Another sport with eleven players per team, **Cricket**, where teams may select a "**12th** man", who may replace an injured player for the purpose of fielding (but not batting or bowling).

A **Quarter** lasts **12** minutes in an **NBA** game.

In Technology

On most **PC keyboards**, the number of function keys are numbered **F1** through F**12**.

In all standard digital telephones, the number of keys are **twelve** (**1** through **9, 0, *** and **#**).

Chapter 8: <u>The Number 12 in Film, Television, Theater, & Literature</u>

In the Arts

Film

Movies with the number **twelve** or its variations in their titles include:

The Dirty Dozen

Ocean's Twelve

12 Monkeys

Cheaper by the Dozen

12 Angry Men

12

Twelve

12 Rounds

Twelve Years a Slave

The Twelve Chairs

Television

In the television franchise **Battlestar Galactica**, the characters come from the **Twelve Colonies of Kobol** and worship the **twelve** lords there. There are also **twelve** models of the humanoid version of **Cylons** in the re-imagined series.

The **Star Palace** is home to the **twelve Star Princesses** in **Star Twinkle PreCure**, one for each sign of the **Zodiac**.

A group of **American** regional cable news television channels covering **New York, New Jersey, and Connecticut** is known as the **News 12 Networks**.

An episode of the television show **The Twilight Zone** was "**Number 12 Looks Just Like You**".

An animated television show on **Adult Swim** was **12** oz. Mouse.

A **French** game show broadcast on **TF1** with **Oliver Minne** at midday **CEST** is **Les 12 Coups de Midi**.

An alien child using base-**twelve** arithmetic in the short "**Little Twelvetoes**" was portrayed on **Schoolhouse Rock!**

The original **1954** live performance on the anthology television series **Studio One** was **Twelve Angry Men**.

Theatre

Adapted from his own teleplay (see above), was **Twelve Angry Men by Reginald Rose**.

A **Jacobean** masque by **Samuel Daniel** was **The Vision of the Twelve Goddesses**.

And of course, the comedy by **William Shakespeare**: **Twelfth Night**.

Literature

A **1946** novel by **Frank Bunker Gilbreth, Jr. and Ernestine Gilbreth Carey**: **Cheaper by the Dozen**.

The epic poem: **Aeneid** by **Virgil** is divided into two halves composed of **twelve** books.

The epic poem: **Paradise Lost** by **John Milton** is divided into **twelve** books perhaps in imitation of the **Aeneid**.

Aleksandr Blok wrote the poem: The **Twelve**.

Twelve Novelas Ejemplares was written by **Miguel de Cervantes**.

Nick McDonell wrote a novel entitled: **Twelve**.

There is a folk tale called: The **Twelve Dancing Princesses**.

A satirical novel by the **Soviet** authors **Ilf and Petrov** was titled: **The Twelve Chairs**.

Chapter 9: <u>The Number 12 in Music, Art, & Games</u>

<u>Music</u>

<u>Music theory</u>

One of the most prominent chord progressions in popular music is the **Twelve bar blues**.

A **78 rpm** phonograph record is **12** inches in diameter.

Devised by **Arnold Schoenberg**, the **twelve-tone technique** (also known as **dodecaphony**) is a method of musical composition.

The interval of an octave and a fifth is known as a **twelfth**. A sopped cylindrical pipe instrument, such as the clarinet, overblow at the **twelfth**.

The number of pitch classes in an octave, not counting the duplicated (octave) pitch is **twelve**. The total number of major keys, (not counting enharmonic equivalents) and the total number of minor keys (also not counting equivalents) is **twelve**, applying only to **twelve** tone equal temperament, the most common tuning used today in western influenced music.

<u>Pop music</u>

A progressive rock band: **Twelfth Night**.

An R. Kelly album: **12 Play**.

A mathcore band: The **Number 12 Looks Like You**.

A song from the album **Brave Murder Day by Katatonia**: **12**.

An album by **Patti Smith**: **Twelve**.

A rap group also known as the **Dirty Dozen**: **D12**.

An **American Christian** rock band: **12 Stones**.

An all-female **Chinese** musical group: **Twelve Girls Band**.

A song by **American** rock band **Marilyn Manson**: **Revelation #12**.

The **12th** studio album by **Keller Williams**: **12**.

A studio album by **German** singer **Herbert Grönemeyer**: **12**.

An album by **Italian** singer **Mina**: **12 (American Song Book)**.

One of the most well-known hits by **Anna Vissi**: **12**.

A song by band **Mushroomhead** of their **Savior Sorrow album**: **12 Hundred**.

An album by **Cyndi Lauper**: **Twelve Deadly Cyns...and Then Some**.

The gift on the **twelfth** day of **Christmas** in the carol 'The **Twelve** Days of **Christmas**: **Twelve** drummers drumming.

A vinyl record format: **12**-inch single.

Art theory

Twelve is the number of basic hues in the color wheel; 3 primary colors, 3 secondary colors, and 6 tertiary colors.

Games

A character in the **Street Fighter** video game series: **Twelve**.

The starting grid in both, **Mario Kart 8** and **Mario Kart Wii** carries **twelve** characters in each race.

Having a long history of **12** points on each side of the gaming board, as evidenced in the **XII** scripta board in the museum at **Ephesus**, is **Backgammon**.

A dice roll of two sixes (value **12**) on the come-out roll constitutes a 'craps' and the shooter (dice thrower) loses immediately in the game of craps.

Chapter 10: <u>The Number 12 in Religion</u>

Religion

The number **twelve** carries religious, mythological and magical symbolism, generally representing perfection, entirety, or cosmic order in traditions since antiquity.

Ancient Greek Religion

In history, a dodecapolis is several sets of **twelve** cities, with the most familiar being the **Etruscan League**. **Twelve** lictors carried fasces of **twelve** rods in ancient **Rome**.

Preceded by **twelve Titans**, were the **twelve Olympians**, the principle gods of the pantheon. Then, there are the **twelve** labors of **Hercules**.

Judaism and Christianity

Twelve Great Feasts are observed by the **Eastern Orthodoxy**.

The first-born son of **Abraham, Ishmael**, had **12** sons/princes (**Genesis 25:16**).

The progenitors of the **Twelve Tribes of Israel** were the **12** sons of **Jacob**.

A meeting is held to add **Saint Matthias** to complete the apostles to number **twelve** once again, after **Judas Iscariot** is disgraced and hangs himself.

In the **Book of Revelation (12:1)**, mentions a woman, interpreted as the people of **Israel**, the **Church**, and the **Virgin Mary**, wearing a crown of **twelve** stars (representing each of the **twelve** tribes of **Israel**).

Also, in the **Book of Revelation** there are **12,000** people sealed from each of the **twelve** tribes of **Israel**, making a total of **144,000** (the square of **12** multiplied by **1000**).

The interval between **Christmas** and the **Epiphany** is counted by the 'Twelve Days of Christmas'.

Hinduism

Frequently, it is said that there are **12 Adityas**.

In **Anahata** or 'heart chakra' there are **12 Petals**.

The **Monkey god Hanuman** has **12** names, as does the **Sun god Surya**.

In **Hindu** temples across **India**, according to the **Shaiva** tradition, there are **twelve Jyotirlinga (Self-formed Lingas)** of **Lord Shiva**.

Chapter 11: The Number 12 in Miscellaneous Other Areas

Misc. Others

Odin, the chief **Norse** god had **twelve** sons.

There were **12** gods of **Olympus** in **Greek** mythology.

The ancient legislation underlying **Roman law** was the **Twelve Tables** or **Leges Duodecim Tabularum**, more informally known as the **Duodecim Tabulae.**

The **French** department, **Aveyron**, the **INSEE and Post Code** is **12**.

There are **twelve Imams**, legitimate successors of the **Islamic** prophet, **Muhammad**, in **Twelver Shi'a Islam**. These **twelve** early leaders of **Islam** include: **Husayn**, nine of **Husayn's** descendants, **Ali, and Hasan**.

Sura Yusuf , narrating the story of the sons of **Jacob**, in the **Quran** is **Sura 12**.

There were **12 Knights of the Round Table**.

Arthur, in the **King Arthur Legend**, is said to subdue **12** rebel princes and to win **12** great battles against **Saxon** invaders.

A **dozen** is **12** of something. A **Baker's dozen**, is when the baker throws in a freebie for you buying the **dozen**, making it 13.

In a **troy pound** (used for precious metals), there are **12 troy ounces**.

The **Flag of Europe** features **12** stars.

There were **twelve** pence in a shilling in the former **British** currency system.

Twelve is the number of greatest magnitude that has just one syllable in the English language, and also the last one to contain a single syllable.

One typically attends school through the **12th** grade.

In the human body, there are normally **twelve** pairs of ribs.

A list of **12** biochemical cell salts, also known as tissue salts was developed by **Wilhelm Heinrich Schüßler**.

There are **12** steps, **12** traditions and **12** concepts for world service in **Alcoholics Anonymous**.

The number of people to have walked on the **Earth's moon** is **twelve**.

Majestic 12 is a secret committee set up by **President Harry S. Truman** to investigate the **Roswell UFO incident** and continue to cover up future extraterrestrial contact.

Chapter 12: <u>The Conclusion</u>

As you have seen by the many examples in this book, the **number 12** is everywhere. It is a very important number to the human race, and is believed by some to have been given to us by the extra-terrestrial aliens who first inhabited this planet.

So, whenever you are faced with the number **12**, be aware of its importance and how special it really is. Whenever possible, attempt to incorporate it into your life. If you have a business that has 10 or 11 employees, hiring **number 12** just might make a huge difference in your business.

If you have 11 suits, pairs of shoes, shirts, etc., purchasing the **12th** one just might make a huge difference in your life.

It should be noted here, that the number **12** occurs **288** times in this book (including the word **dozen** (this is a product of **12** x 24). Hopefully, using this **Divine Number** will help promote this book, as well as, others that I have written.

<u>Best of Luck in All Things in this Life!</u>

Sincerely,

CJ Dodaro

About the Author

CJ Dodaro

I am a 68 year young man, who has worked very hard my whole life. Although I have occasionally worked for other companies, I have been an entrepreneur since I was a young boy.

I starting out selling things door to door when I was around 10 years old; seeds, greeting cards, and wrapping paper for all occasions. After a couple of years, I graduated to lemonade/Kool-Aid stands, washing cars and cutting neighbor's lawns. I did just about anything to make a few bucks.

When I turned 21 years old, I had the opportunity to purchase a portion of an existing landscape maintenance business from the gentleman I had been working for. I really enjoyed the work and soon built it into a full service maintenance and installation company which I operated successfully for over 25 years, before passing it on to my son.

I have also owned and operated a soup and sandwich restaurant, a gas station, and a body shop. I also had a three-piece 50s & 60s rock & roll band which was a lot of fun for many years.

Also, being a homeowner for over 30 years, with usually more time on my hands than money, I learned every aspect of home building and remodeling; from pouring a foundation to putting on a new roof and everything in between. I have installed tile, hardwood floors, carpeting, kitchen cabinets, toilets, vanities, etc. I have learned to do electrical work and plumbing as well.

Forced to retire at the age of 59 due to a severe back injury during a construction project, I found myself in unfamiliar territory; lots of time on my hands, but physically limited. So, after about 6 months of physical therapy, spending some time organizing things around the house, and doing some light projects, I began to go a little stir crazy. So, I decided to sit down and teach myself the computer.

I caught on fairly quickly and was soon operating a couple of e-commerce websites, selling **DVDs, CDs**, and miscellaneous household items. I did that for a couple of years and then began to learn about the **Affiliate Marketing Business**; another new adventure. I operated (2) different websites for that business for a couple of years, which I have since shut down, realizing that I could do more by **Paying-it-Forward**. So, I decided to teach myself how to **write and self-publish eBooks** so that I could share the knowledge that I had acquired from starting and running several of my own successful businesses and **Improving America LLC** was born.

Our **website** is: **https://improvingamericallc.com/**. You should definitely check it out.

Improving America LLC

Improving America LLC is a multi-faceted company whose main objective is to help entrepreneurs create and/or improve their businesses. We do this in the following ways:

• We create **step-by-step eBooks and Paperbacks** about how to create and run a successful business, along with some books on specific types of businesses and projects.

• We are an **Angel Investor** in new business startups, helping deserving companies build their businesses into successful ones.

• We also invest in **Tax Lien Certificates and Tax Deeds** by purchasing these delinquent taxes in order to help **Counties** across the **USA** obtain some much needed funds for the services that they provide their residents, such as their: **Fire Depts., Police Dept., libraries, schools, etc.**

• We resell the properties acquired through unredeemed **Tax Lien Certificates and Tax Deeds** to **Rehabbers and house Flippers**, at a tremendous discount, in order to provide jobs, along with additional housing for interested home owners across the country.

• We help people recover **Unclaimed Property** that they have either forgotten about or didn't know existed.

• We assist people who have gone through a **Tax Sale Foreclosure** recover **Overages** that they are due from the sale.

• We **Fix & Flip properties**, as well as, **Fix & Hold others**. We provide work for **Contractors** and excellent housing potentials for interested **Buyers** and **Renters**.

Those of you that are interested in becoming entrepreneurs, with a strong desire to be your own boss and own your own business, or you are already an entrepreneur who needs advice on improving your business, please click on the **My eBooks** tab in the **Main Menu** at the top, and explore the many helps that I provide.

If you own a budding company that is looking for funds to increase their business, please send me a **mini-business plan** showing me what you and your company are made of, along with the direction that you intend to take, and I will be happy to review it. If I think that it is a solid plan, I will gladly point you in the direction to obtain some funding for your project and I may even invest in you myself. The **mini-business plan** should be limited to no more than 15 pages, showing me the highlights of you and your business. Send your email to: **cj@improvingamericallc.com**.

It is my intention to continue writing more books, specifically geared toward starting and operating specific businesses and completing specific projects. You can be kept apprised of my current projects, by signing up for my **Email notifications** on my **website**: **https://improvingamericallc.com/** or you may personally **Email** me at: **cj@improvingamericallc.com** in order to ask specific questions.

Best of Luck in Life and All You Do!

Sincerely,

CJ Dodaro

Improving America LLC

Other Books by This Author

Thank you for reading my book. Please take a moment to leave me a Review at your favorite retailer.

CJ Dodaro - Author

You can click on each of the **individual book links** to be directed to view a **20% sample of the book**. My **Other Books** are:

Being Successful in Your Own Business - A Step-by-Step Guide to Success

A (3) Book Series:

Book 1 of 3 in the **Series:** **Your Business Setup**

(See: **https://amzn.to/3lth67d**)

You will learn how to choose: your Business, your Business Name, and your Business Structure. You will also learn how to: set up a Home Office, set up your Bookkeeping, create a Business Plan, get your Business Licenses, choose and purchase your Domain Name, choose a Host, choose a Website Theme, and initially set your site up.

Book 2 of 3 in the **Series:** **Work on Your Website**

(See: **https://amzn.to/3fUThUN**)

You will learn how to create: a Logo, a Home Page, additional Pages, Menus, Sidebar Widgets, and a Blog Post. You will also learn about: Keywords, Images, Affiliate Links, Product Reviews, and how to best benefit from Google and Bing.

Book 3 of 3 in the **Series:** **Get Social, Videos, and Money Management**

(See: **https://amzn.to/3qilQ30**)

You will learn how to set up and work with: Facebook, Twitter, and Pinterest. You will also learn: how to create Videos, about Money Management, how to do your own Taxes, and how to create Additional Income Streams.

Book 4: **Write it - Publish it - FREE!**

(See: **https://amzn.to/3fYEXdD**)

Instructions for Writing, Formatting, Publishing, and Marketing an eBook and a Paperback.

Book 5: <u>Make Money - Work at Home with an Unclaimed Money Recovery Business</u>

(See: <u>https://amzn.to/3mvsLUe</u>)

An <u>Unclaimed Money Recovery Business</u> is the perfect business to run out of your home. If you are looking for a way to supplement your income, which can eventually turn into a full-time career, then this could be for you.

Book 6: <u>Make Money - Work at Home with a Tax Sale Overages Business</u>

(See: <u>https://amzn.to/33Aj0g3</u>)

A <u>Tax Sale Overages Business</u> is the perfect business to run out of your home. If you are looking for a way to supplement your income, which can eventually turn into a full-time career, then this could be for you.

Book 7: <u>Make Money - Work at Home with a Tax Lien Certificates & Tax Deeds Business</u>

(See: <u>https://amzn.to/3ohwghs</u>)

A Tax Lien Certificates & Tax Deeds Business is the perfect business to run out of your home. If you are looking for a way to supplement your income, which can eventually turn into a full-time career, then this could be for you.

Book 8: <u>How to Build a Deck - Step-by-Step Instructions</u>

(See: <u>https://amzn.to/3qj287r</u>)

This book is a step-by-step guide which will walk you through the building of your deck in a simple, easy to follow manner.

Book 9: <u>The Right Way to Flip Houses A Step-by-Step Guide to Success</u>

(See: <u>https://amzn.to/39ZJav8</u>)

This book is a step-by-step guide which will teach you the Right Way to Flip Houses. If you follow my guidance and apply everything as I have laid it out, you will soon be on your way to a very lucrative career.

Book 10: <u>Doing Your Due Diligence - The Companion Book for: The Right Way to Flip Houses</u>

(See: <u>https://amzn.to/3qR0Pfu</u>)

This book is the Companion Book for: The Right Way to Flip Houses. It is a Step-by-Step Guide for the Due Diligence processes that need to take place for this business. Following the Steps in this book are crucial for your Real Estate success.

<u>**Book 11**</u>: <u>**Make Money - Work at Home**</u> with an <u>**Affiliate Marketing Business**</u>

(See: **https://amzn.to/3j07z88**)

This book is a **Step-by-Step set of instructions** which will teach you how to <u>**Make Money - Work at Home with an Affiliate Marketing Business**</u>. This is a business that you can start part-time and build into a successful full-time career.

There will more **eBooks** coming in the near future. You can be kept apprised of their creation, by visiting my website: **https://improvingamericallc.com/** or you may personally **Email** me at: **cjdodaro@att.net** in order to ask specific questions.

Also, all of my **eBooks**, as well as **Paperback versions** of each one are available on my <u>**Author's page on Amazon KDP**</u> at: (**https://amzn.to/363a6JB**).

If you prefer to see a **sample of any of my books**, click on my <u>**Author's page on Smashwords.com**</u>. (See: **https://www.smashwords.com/profile/view/CJDodaro?ref=CJDodaro**). Once there, you can **Scroll** down to see all of my **Published books** and click on the one that you wish to view.

<u>Best of Luck in Your Business and in All That You Do!</u>

Sincerely,

CJ Dodaro

Contact Me

There are several ways to **Contact Me and/or View or Purchase My Books**. These ways are:

Contact Me

1. On my website: **https://improvingamericallc.com/**.

   Contact me directly at:**cj@improvingamericallc.com**

2. You may personally **Email me at**: **cjdodaro@att.net** in order to ask any specific questions.

View or Purchase My Books

1. **Smashwords.com**

   • See my **Author's page on Smashwords.com** at: **https://www.smashwords.com/profile/view/CJDodaro?ref=CJDodaro**.

   • **Scroll** down to see all of my **Published books** and click on the one that you wish to view. You can **download a FREE 20% Sample** to review each one before you make your purchase.

2. **Amazon Kindle Direct**

   • See my **Author's page on Amazon KDP** at: **https://amzn.to/363a6JB**.

   • You can purchase any of my books in the **eBook and/or Paperback version**.

3. **My Website**: **https://improvingamericallc.com/**.

   • Click on the: **'eBooks' tab** (on top). You will find a list of all of my **eBooks (and Paperback versions)**, along with a description for each one.

   • You can purchase any of my books in the **eBook and/or Paperback version** through a secured payment method.

   • You can also leave me a **Review** for any of my books by clicking on the: **'eBooks' tab** (on top) and leaving the **Review** on the appropriate **Post**. I thank you in advance.

There will more **eBooks** coming in the near future. You can be kept apprised of their creation, by visiting my website: **https://improvingamericallc.com/**. Click on the: **'eBooks' tab** (on top) to view **Posted information**.

Thank you.

Sincerely,

CJ Dodaro - Author